W9-BYE-057

Our Universe

Uranus

by Margaret J. Goldstein

Lerner Publications Company • Minneapolis

Lerner Publications Company
A division of Lerner Publishing Group
241 First Avenue North
Minneapolis, MN 55401 USA

Website address: www.lernerbooks.com

Words in **bold type** are explained in a glossary on page 30.

Library of Congress Cataloging-in-Publication Data

Goldstein, Margaret J.
Uranus / by Margaret J. Goldstein.
p. cm. – (Our universe)
Includes index.
Summary: An introduction to Uranus, describing its place in the solar system, its physical characteristics, its movement in space, and other facts about this planet.
ISBN: 0–8225–4654–X (lib. bdg. : alk. paper)
1. Uranus (Planet)–Juvenile literature. [1. Uranus (Planet)] I. Title. II. Series.
QB681 .G67 2003
523.47–dc21

Manufactured in the United States of America
1 2 3 4 5 6 – JR – 08 07 06 05 04 03

This giant planet looks like a bright blue-green ball. It has rings and many moons. Do you know its name?

The planet is called Uranus. Uranus is a large and faraway planet. It is much larger than our planet. More than 60 planets the size of Earth could fit inside Uranus.

Uranus and Earth are part of the **solar system.** The solar system has nine planets in all. The Sun is at the center of the solar system.

All of the planets in the solar system **orbit** the Sun. To orbit the Sun means to travel around it. Uranus orbits the Sun in an oval path. It is the seventh planet from the Sun.

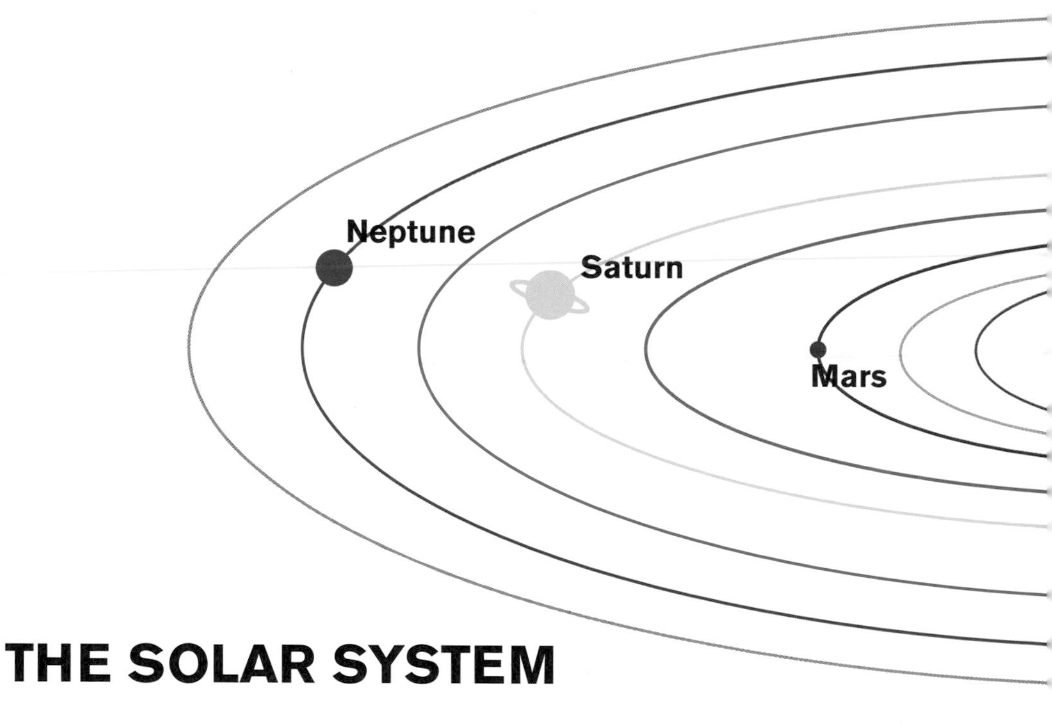

Uranus makes a long trip around the Sun. It takes about 84 years to orbit all the way around once. Earth takes only 1 year to orbit the Sun.

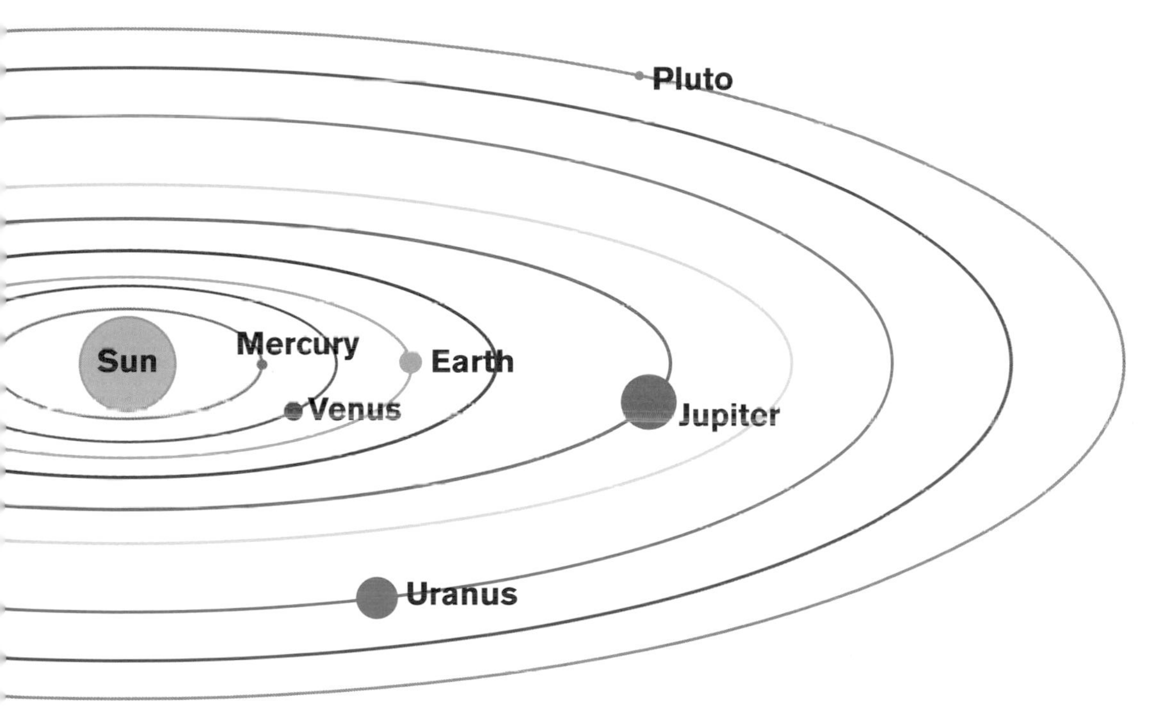

Uranus moves in another way, too. It spins around like a top. This kind of spinning is called **rotating.** Uranus rotates all the way around in about 17 hours. All of the planets rotate. Earth takes 24 hours to rotate once.

Uranus rotates on its **axis.** An axis is an imaginary line that goes through the center of a planet. Most planets have a tilted axis.

Uranus's axis is tilted very far. That means Uranus rotates on its side. Many years ago, another planet may have crashed into Uranus and knocked it sideways.

URANUS'S AXIS

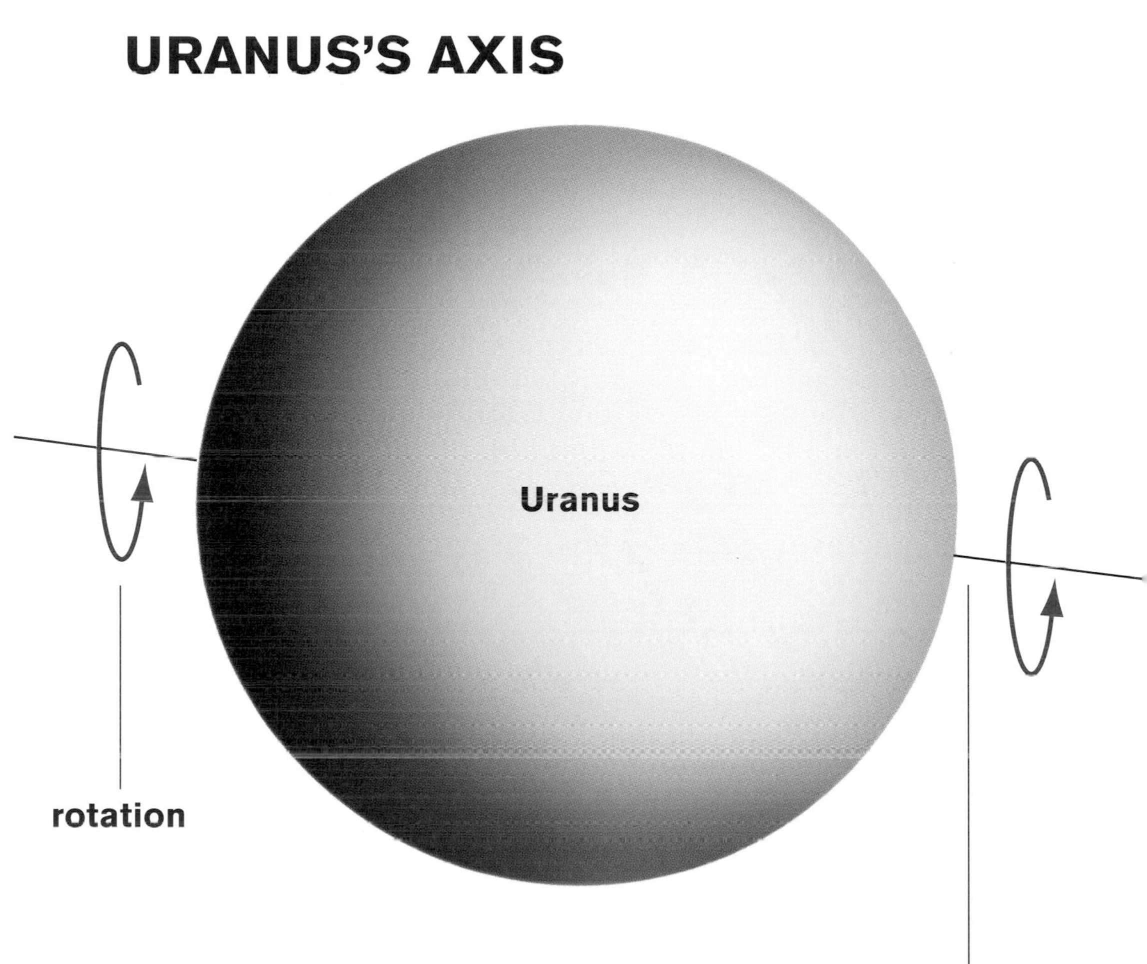

Uranus does not have hard, solid ground like Earth has. Uranus is made mostly of gases and liquids. A thick layer of gases surrounds the planet. This layer is called an **atmosphere.**

Below the atmosphere is probably a deep icy ocean. The very center of Uranus may be solid ice and rock.

URANUS'S LAYERS

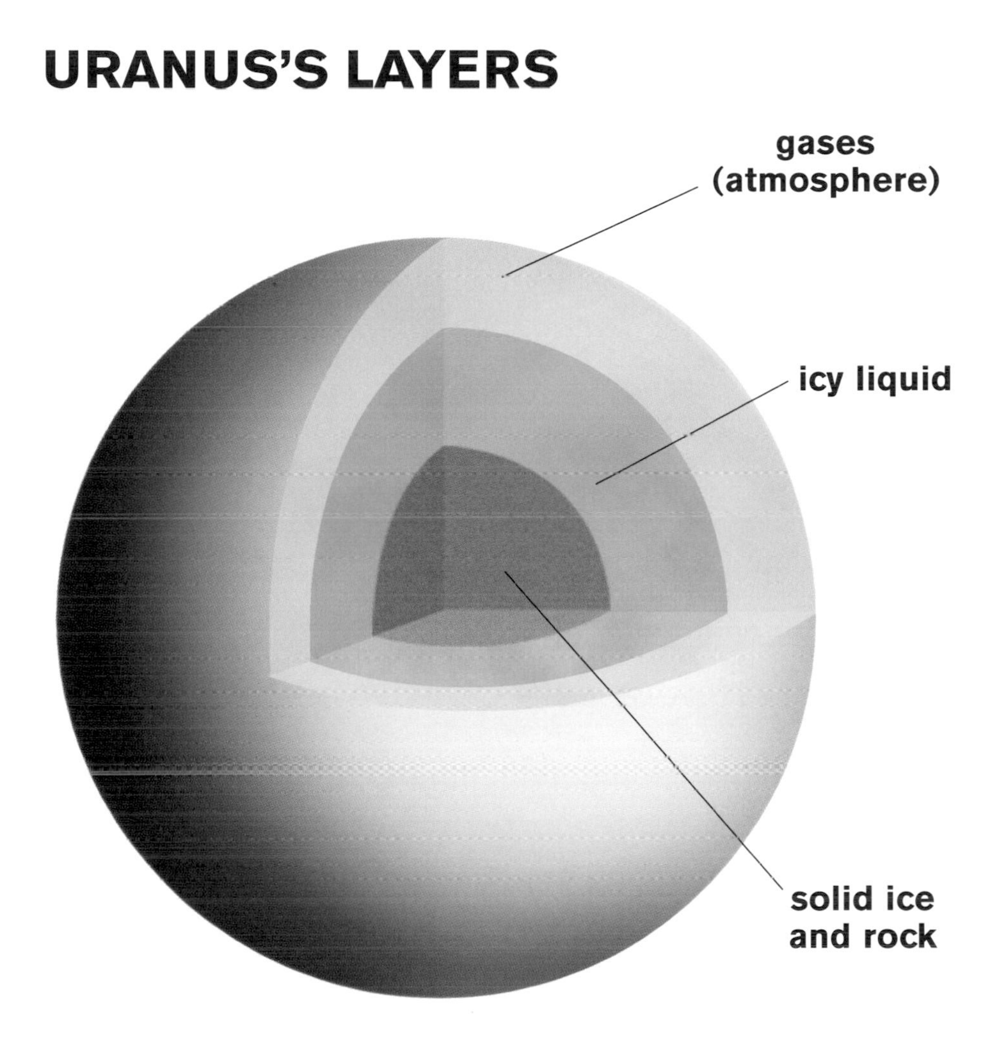

Uranus is far away from the Sun's heat. So its atmosphere is very cold. Bright blue-green clouds float through the cold atmosphere. These clouds are made of a gas called methane. The methane gas is frozen into ice.

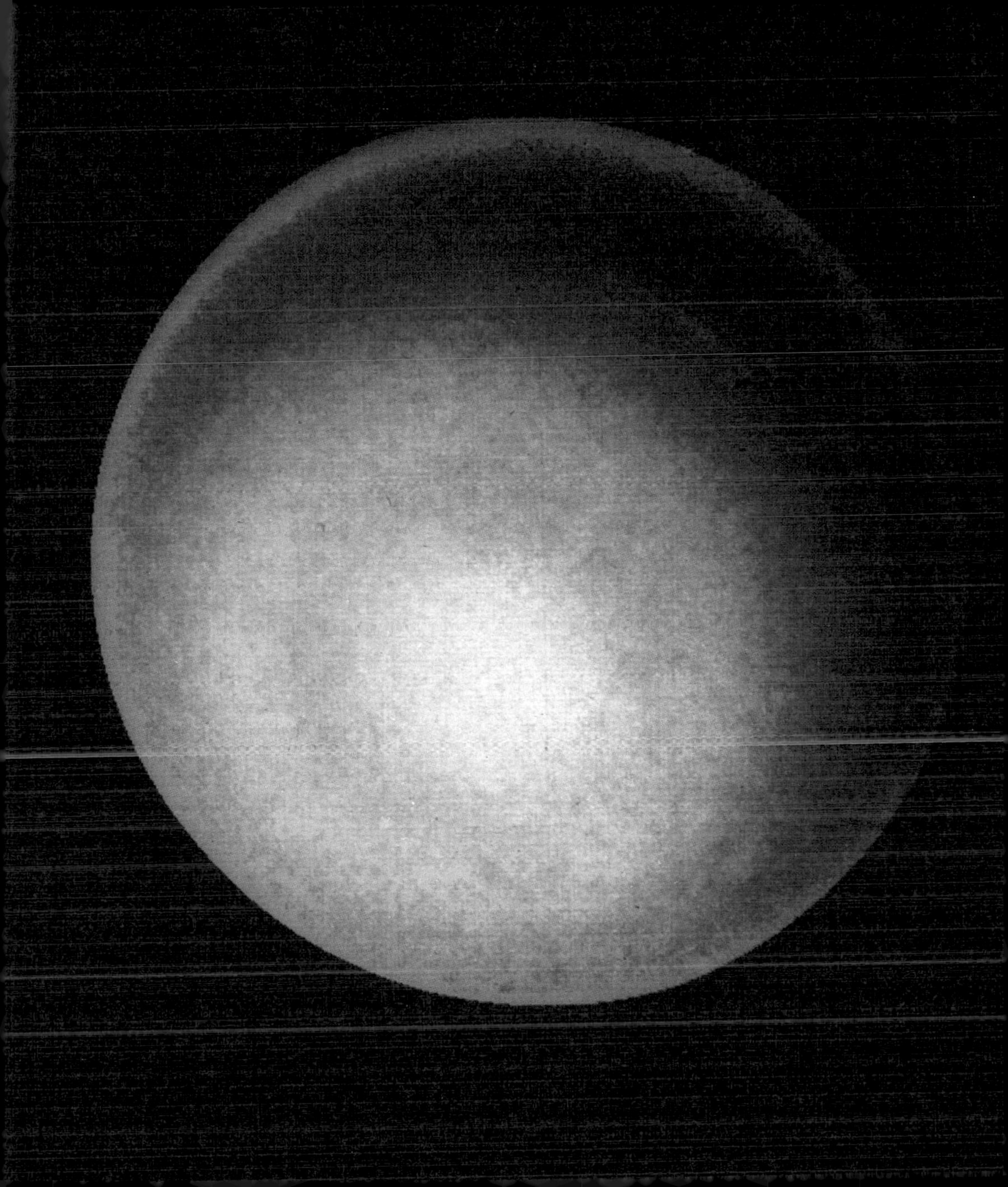

An artist made this illustration of Uranus's rings.

Something special orbits Uranus. Eleven rings circle the planet. Only three other planets in the solar system have rings. These planets are Saturn, Jupiter, and Neptune.

Uranus's rings are made of little pieces of ice. The rings are hard to see. They are very thin. And they are not as bright as Uranus is.

Uranus has some nearby neighbors in space. They are its moons. The moons are much smaller than Uranus is. They orbit the planet the way Uranus orbits the Sun.

Uranus has at least 21 moons. The five biggest moons are named Ariel, Umbriel, Titania, Oberon, and Miranda.

Oberon

Uranus's largest moons are round, rocky, and icy. They are all covered with deep holes called **craters.**

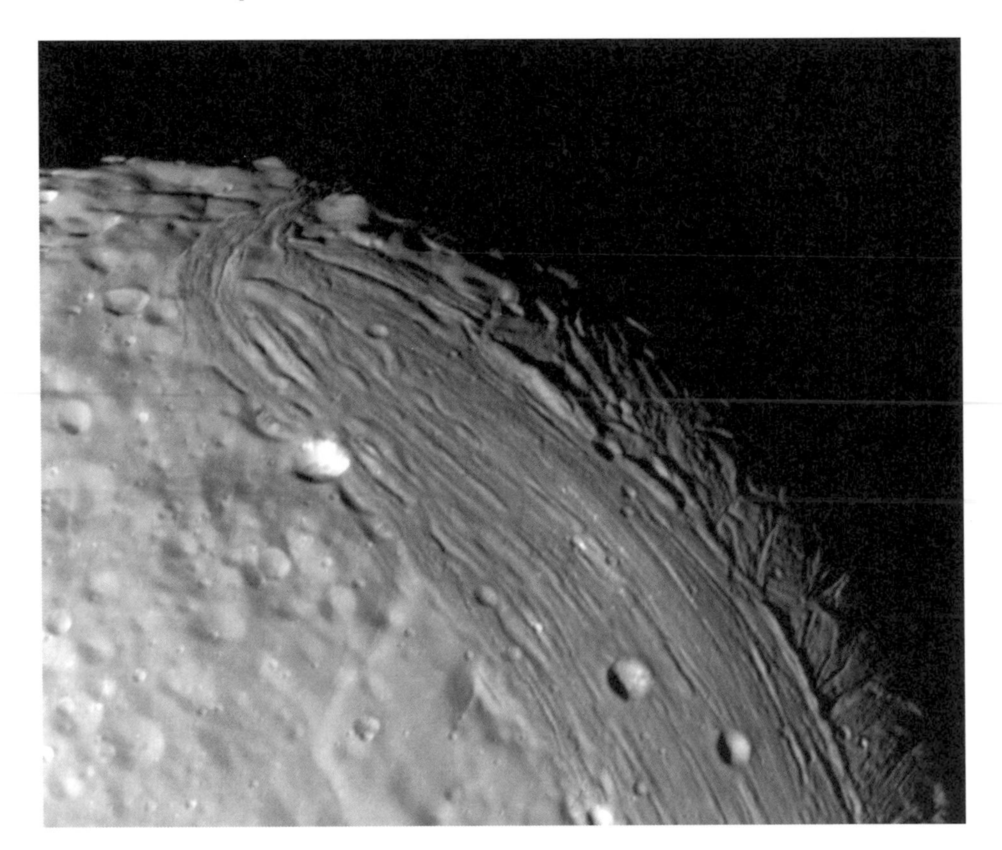

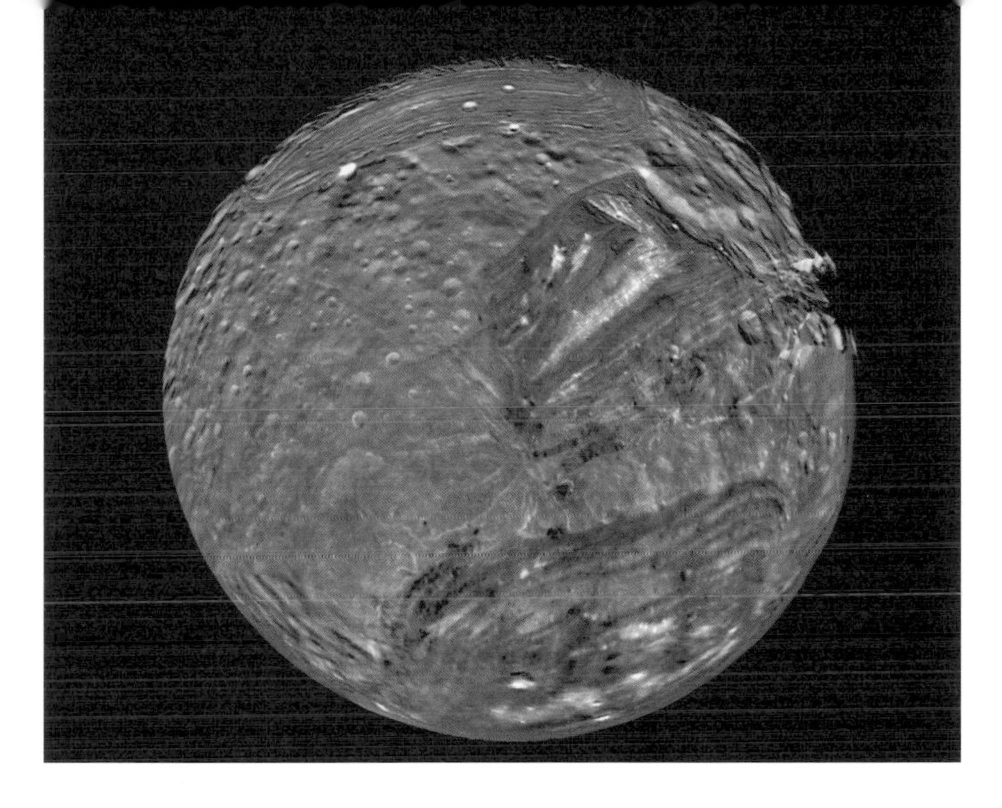

Miranda looks different from Uranus's other moons. It has deep and jagged valleys. It has crisscrossing cliffs. And it has a set of curving grooves. The grooves look like a large racetrack!

An **astronomer** named William Herschel discovered Uranus in 1781. An astronomer is someone who studies outer space. Herschel saw Uranus through a telescope.

Herschel later discovered the moons Titania and Oberon. But he did not see the other moons. And he did not see the rings.

Astronomers kept studying Uranus in the night sky. They found many more moons. In 1977, they learned that Uranus has rings. But they still had much more to learn about the planet.

A spacecraft visited Uranus in 1986. It was called *Voyager 2*. *Voyager 2* carried cameras and other machines. It gave astronomers a chance to see Uranus up close.

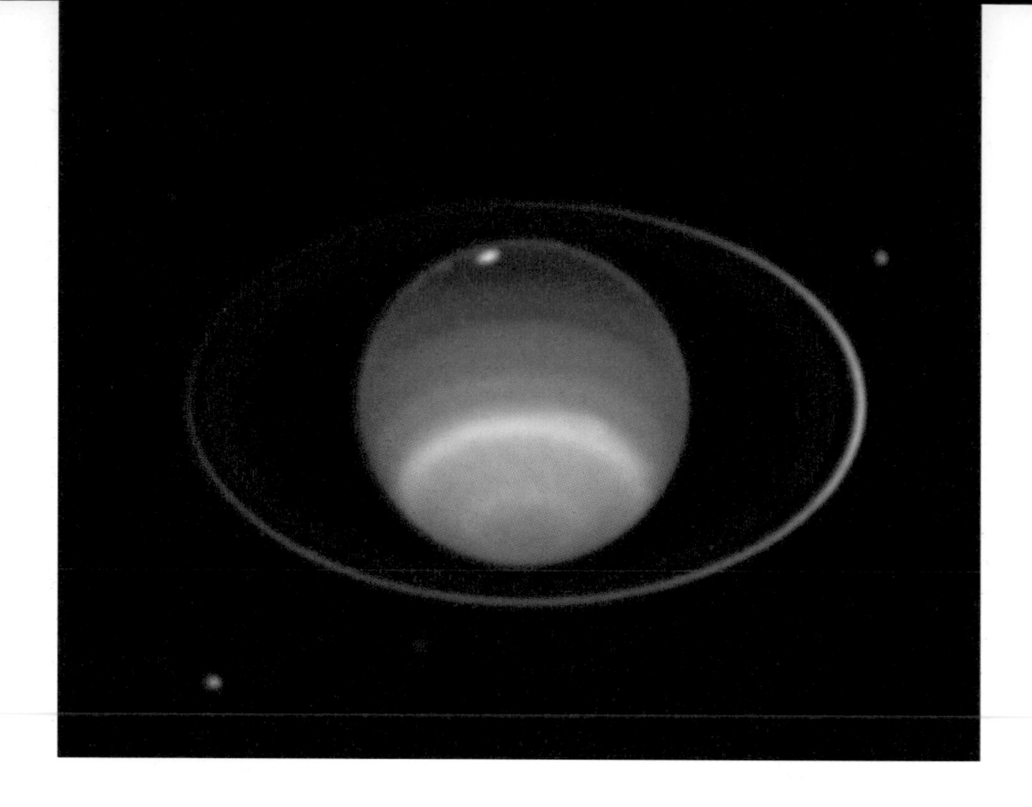

Voyager 2 took pictures of Uranus. It studied the atmosphere. And it explored the planet's rings and moons. The spacecraft found 10 moons that no one had known about.

Imagine that you could visit Uranus.
What do you think you might find?

Facts about Uranus

- Uranus is 1,780,000,000 miles (2,870,000,000 km) from the Sun.
- Uranus's diameter (distance across) is 31,800 miles (51,100 km).
- Uranus orbits the Sun in 84 years.
- Uranus rotates in 17 hours.
- The average temperature in Uranus's atmosphere is −323°F (−197°C).
- Uranus's atmosphere is made of hydrogen, helium, and methane.
- Uranus has 21 moons.
- Uranus has 11 rings.
- Uranus was discovered in 1871 by William Herschel.

- Uranus was named after the Greek god of the heavens.
- Uranus was visited by *Voyager 2* in 1986.
- Uranus and Neptune are almost exactly the same size. Uranus is just a little bigger than Neptune.
- Winds on Uranus can blow up to 450 miles (725 km) per hour.
- Uranus is called a "gas giant" because it is very big and is made mostly of gas.
- Many of Uranus's moons are named after characters from British literature.
- Uranus was the first planet to be discovered in modern times. People had known about Mercury, Venus, Mars, Jupiter, and Saturn for thousands of years.

Glossary

astronomer: a person who studies outer space

atmosphere: the layer of gases that surrounds a planet or moon

axis: an imaginary line that goes through the center of a planet

craters: large holes on a planet or moon

orbit: to travel around a larger object in space

rotating: spinning around in space

solar system: the Sun and the planets, moons, and other objects that travel around it

Learn More about Uranus

Books

Brimner, Larry Dane. *Uranus.* New York: Children's Press, 1999.

Simon, Seymour. *Uranus.* New York: Morrow, 1990.

Websites

Solar System Exploration: Uranus
<http://solarsystem.nasa.gov/features/planets/uranus/uranus.html>
Detailed information from the National Aeronautics and Space Administration (NASA) about Uranus, with good links to other helpful websites.

The Space Place
<http://spaceplace.jpl.nasa.gov>
An astronomy website for kids developed by NASA's Jet Propulsion Laboratory.

StarChild
<http://starchild.gsfc.nasa.gov/docs/StarChild/StarChild.html>
An online learning center for young astronomers, sponsored by NASA.

Index